Chemical Engineering & the Law Forum 2019

Held at the 2019 AIChE Annual Meeting

Orlando, Florida, USA
10-15 November 2019

ISBN: 978-1-7138-0545-8

Printed from e-media with permission by:

Curran Associates, Inc.
57 Morehouse Lane
Red Hook, NY 12571

Some format issues inherent in the e-media version may also appear in this print version.

Printed with permission by Curran Associates, Inc. (2020)

For permission requests, please contact AIChE
at the address below.

AIChE
120 Wall Street, FL 23
New York, NY 10005-4020

Phone: (800) 242-4363
Fax: (203) 775-5177

www.aiche.org

Additional copies of this publication are available from:

Curran Associates, Inc.
57 Morehouse Lane
Red Hook, NY 12571 USA
Phone: 845-758-0400
Fax: 845-758-2633
Email: curran@proceedings.com
Web: www.proceedings.com

TABLE OF CONTENTS

(148A) ASSERTING AND PROTECTING IP RIGHTS IN CHEMICAL AND PHARMACEUTICAL INDUSTRIES 1
Hana Oh

(148B) PATENTS ARE INTEGRAL TO MONETIZING INNOVATION. COURT DECISIONS HAVE LED TO UNCERTAINTY AS TO WHAT IS PATENTABLE, AFFECTING DRUG, DIAGNOSTIC, GENETIC, AND OTHER INNOVATION 2
Matthew Zapadka

(148C) START-UP, SPIN-OUT, DON'T BURN-OUT: PROTECTING YOUR INTELLECTUAL ASSETS ON A SHOESTRING BUDGET 4
Robert Sovesky

(148D) HOW TO PROTECT YOUR INTELLECTUAL PROPERTY IN THE ACADEMIC WORLD 5
Paul Townsend, Charles Collins-Chase, Lauren Dowty

(148E) LIFE CYCLE MANAGEMENT FOR LIFE SCIENCE INVENTIONS 6
Shan Liu

(508A) FINDINGS FROM MTO COMMERCIALIZATION – DON'T FORGET ABOUT THE SMALL STUFF 7
Joe Montalbano

(508B) DE-RISKING TECHNOLOGY DEVELOPMENT AND SCALE-UP EFFORTS EFFECTIVELY IN FUEL AND CHEMICAL PROCESSES 8
Greg DiCosola, Dan Parker, Matt Schneider, Bernardo Flores

(508C) CAN OUTSOURCING R&D BALANCE MYOPIC BUSINESS MANAGEMENT? 9
Michael Bradford

(508D) SUCCEEDING AT INNOVATION: GENERATING VALUE FROM NEW TECHNOLOGY AT EACH STAGE OF DEVELOPMENT AND IMPLEMENTATION 10
Howard Stamato

(508E) HEAT TRANSFER COEFFICIENT IN TWO PHASE LIQUID-SOLID FLUIDIZED BED 11
Jamal M. Ali

Author Index

578382 Asserting and Protecting IP Rights in Chemical and Pharmaceutical Industries

Obtaining and defending technological innovations is a critical issue for scientists, academics, and executives to understand. Due to developments in technology and in law, the interplay between different types of intellectual property ("IP") rights are evolving, and not always mutually exclusive as traditionally viewed. I will discuss practical approaches for different forms of IP, using real-world examples and cases, with particular focus on patents and trade secrets. We will discuss recent developments, such as the trade secret cases brought in Federal Court under the 2016 Federal trade secret statute, updates to patent-eligible subject matter, and cases involving multiple forms of IP -- and how they may affect the chemical and pharmaceutical industries.

568355 Patents Are Integral to Monetizing Innovation. Court Decisions Have Led to Uncertainty As to What Is Patentable, Affecting Drug, Diagnostic, Genetic, and Other Innovation

It has been nearly five years since the U.S. Supreme Court issued the *Alice Corp. Pty. Ltd. v. CLS Bank Intern.* decision and about seven years since its decision in *Mayo Collaborative Servs. v. Prometheus Labs., Inc.*. Those decisions established the current framework—dubbed the "*Alice/Mayo* test"—for determining the subject matter (*i.e.*, technology) eligible for patent protection. Since the creation of the *Alice/Mayo* test, the primary appellate court that handles patent appeals has issued many decisions hinging on this issue, but significant uncertainty remains. This topic is important for those in academia and industry because patents are prevalent and highly-sought after in both arenas.

The *Mayo/Alice* test is a two-step analysis used to interpret § 101 of the Patent Act, which states that a patent can be granted for "any new and useful process, machine, manufacture, or composition of matter." The Supreme Court has created certain exceptions to that rule that include natural products, natural phenomena, and abstract ideas. The Court justified the judicial creation of those exceptions by expressing concern that, without them, a patent could cover the "building blocks" of nature and tie-up their use for future innovation. This is complicated, however, by the fact that all inventions are governed by the natural laws and scientific confines of the universe. When an innovator is working to create new technologies that use or rely upon naturally-occurring materials like DNA, proteins, antibodies, and other chemical constituents, it can be difficult to ascertain whether a court might consider a discovery to fall into one of the exceptions and be ineligible for patent protection.

In general, the *Mayo/Alice* test implicates patents that rely on computers (often characterized as "abstract ideas") or that incorporate or rely on naturally-occurring substances. The latter can be particularly problematic for those working in chemical and biological fields. For example, the *Mayo* case involved a therapeutic method for treating an immune-mediated gastrointestinal disorder that included the steps of administering a drug and then determining the levels of 6-thioguanine after that administration. The Supreme Court declared that therapeutic method was ineligible for patent protection.

Many important technologies have been implicated by this relatively recent jurisprudence. For instance, patent claims for isolated DNA, a method of assessing the risk of a major cardiac event using enzyme levels, a method for detecting mycobacterium tuberculosis, a method of DNA intron sequence analysis, a method of diagnosing a neurological disorder based on the presence of certain autoantibodies, and a prenatal test for certain fetal characteristics using cffDNA were all found to be patent-ineligible natural products or natural phenomena. However, patent claims for treating a patient with schizophrenia, compositions used for increasing athletic performance using beta-alanine, and a process for cryogenically freezing hepatocytes were found patent-eligible. Despite some common technological threads running through these patents, the courts' opinions have made it difficult to cogently predict how similar patents might be treated in the future.

Logically, overbroad patents are often considered to have an inhibitory effect on innovation because they can tie-up later innovators' ability to use preexisting technology to expand upon it and create new technology using it. Much innovation is iterative, after all. However, the U.S. patent system is designed to encourage invention and innovation and when innovators lose the incentive to invest in and create valuable and complex technology because they have no way to protect those inventions, this can be detrimental to innovation and society.

The author of this abstract recently represented a dietary supplement company in a successful appellate reversal of a lower court's ruling that the supplements at issue and methods of treatment using them were patent-ineligible. The patents involved in that case concerned a dietary supplement containing unnatural amounts of beta-alanine that were administered over an extended period of time, resulting in an increase in the anaerobic working capacity of a subject's muscles. That particular method of administration caused an unnatural override of the body's homeostasis. After the lower court concluded those treatments were nothing more than a natural law, the appellate court reversed that decision and held that the patents sufficiently distinguished themselves from a natural law or natural phenomenon and were patent-eligible. Based on this experience, the author intends to discuss the state of patent law, with particular focus on this issue, along with the courts' treatment of the two-step *Mayo/Alice* test. The presentation would also discuss how innovators may seek patent protection on the types of inventions that a court or the U.S. Patent Office might consider subject-matter ineligible.

583619 Start-up, Spin-out, Don't Burn-Out: Protecting Your Intellectual Assets on a Shoestring Budget

The discussion will include a general overview of intellectual property rights and why they are important for emerging growth companies (e.g., entrepreneurs, start-ups, university spin-outs) to consider early in the life of the company. The discussion will compare and contrast patent, trademark, copyright, and trade secret protections. The discussion will include steps an emerging growth company can perform on their own without having to involve an outside consultant (e.g., attorney). The discussion will include common pitfalls when obtaining intellectual property rights and creative ways emerging growth companies can enforce their rights to deter competition or generate revenue. And of course, throughout the discussion, we will discuss how to pay for it all. This discussion is geared toward non-patent professionals, such as engineers, scientists, managers, and c-suite executives.

587156 How to Protect Your Intellectual Property in the Academic World

Universities are full of academics working on cutting edge technology, but how is all that innovation protected? Universities typically have their own organizations in place to help encourage, protect, and license innovations occurring at the school. This presentation will provide an overview of the basics of patent law, how tech transfer works at a university, and important considerations for students, professors, and the university itself. Speakers will also highlight several case studies addressing whether a university's intellectual property protection was sufficient.

588860 Life Cycle Management for Life Science Inventions

The presentation will focus on strategies for building patent portfolio with timing considerations, diversifying patents with alternatives to standard utility filings, and alerting FDA with a citizen petition on a competing product, to extend market exclusivity for a drug or medical device franchise.

579139 Findings from MTO Commercialization – Don't Forget about the Small Stuff

In the rush to commercialize a new technology, there are many things to consider: reactor scale up & design; engineering design of major equipment; simulations; thermodynamics; sales tools; customer discussions, and a lot more. One of the things that can be easily missed is what happens to components that are produced in minor concentrations. It may be easy to ignore these when a simulation indicates that they will be washed away in waste streams (or maybe become product impurities). However, these minor concentration components can become problematic when they don't respond exactly as predicted, especially when they build up their concentrations in recycle loops. This talk will discuss some these type of issues that have occurred with MTO commercialization and what we are doing about them.

573116 De-Risking Technology Development and Scale-up Efforts Effectively in Fuel and Chemical Processes

The sequential pathway for de-risking new technologies includes bench scale testing, pilot programs, demonstration plants and finally commercial plant construction. In the past, hastening, or eliminating any of the above steps would increase the risk to novel technologies, and to the companies that invest in them. Unfortunately, this pathway requires significant time and financial investment. In the past few years, in response to increased pressure to speed up the pathway, engineers have developed many tools and strategies for mitigating that risk. Modern process simulation, targeted piloting programs, toll processors, directional technology development plans and modular strategies can all help companies achieve commercial success from technology development through scale-up to commercialization. When employed properly, these, and other tools can decrease project risk, shorten the time to commercialization, and limit the overall cost of developing innovative technologies. This presentation will discuss how to utilize these, and other tools to quickly and effectively de-risk process technology development. Using case studies of past projects, it will also review the process of developing technologies from bench scale through pilot and demonstration scale to full commercialization within the framework of these de-risking tools. Case studies include next-generation biodiesel development, conversion of natural gas liquids into valuable chemicals and marine fuel, and production of high purity alumina from waste streams.

565316 Can Outsourcing R&D Balance Myopic Business Management?

Abstract

When Ford Motor Co.'s CEO, Jim Hacket, announced plans in April 2018 to reduce spending by $25.5 billion over the next four years [1], Ford staff were concerned about how the pending cuts would impact R&D investments [2]. Similarly, DowDuPont Inc.'s CEO, Ed Breen, announced the elimination of R&D spending on "moonshots" to focus on projects that cost less than $30 million to complete and, perhaps most importantly, carry a lower risk profile [3]. From one perspective, these announcements are merely part of normal operations, as business leaders make decisions to increase, reduce, or cut spending every day. On the other hand, these announcements raise a few questions:

- Were these part of a trend in R&D spending?
- What are some of the factors that affect CEO decision-making?
- What are some consequences of myopic decision-making?
- Can outsourcing R&D balance myopic business management?

After briefly addressing these questions, this presentation will disclose a relevant case study in which MATRIC, a contract R&D organization, successfully helped a client to develop and commercialize a new technology.

References

[1] A.Levine-Weinberg, "Ford Reveals More Details on its Massive Cost-Cutting Plan," Fox Business, August 13, 2018 (https://www.foxbusiness.com/markets/ford-reveals-more-details-on-its-massive-cost-cutting-plan).

[2] L.Brooke, "Underway on Nuclear Power," Automotive Engineering, June 2018, p. 18 (http://www.nxtbook.com/nxtbooks/sae/18AUTP06/index.php#/22).

[3] J.Kaskey, "DowDuPont CEO Kills Off R&D 'Moonshots' in Bid for Higher Profit," Bloomberg, May 31, 2018 (https://www.bloomberg.com/news/articles/2018-05-31/dowdupont-s-new-profit-formula-killing-more-research-projects).

555687 Succeeding at Innovation: Generating Value from New Technology at Each Stage of Development and Implementation

Succeeding at Innovation: Generating Value from New Technology at Each Stage of Development and Innovation

Technology provides increased capability, energizes our imagination, and leads to business advantages. Too often our embrace of technology is weakened by narrow and inaccurate measures of value, the challenge of changing existing practices, and the inertia of our own minds and customs.

Examining the effect new technology has on our systems at each stage of its development and implementation reveals the full value of its effect. The potential gains at each stage, how to map an uncertain path to the future, and methods to decide the level and timing of investment are discussed in the context of drug product delivery, processing, and scale up. Examples from modeling and simulation, 3-D printing, and novel lyophilization technology are used as illustrations.

552804 Heat Transfer Coefficient in Two Phase Liquid-Solid Fluidized Bed

Heat Transfer Coefficient in Two Phase Liquid-Solid Fluidized bed

Marwa Sabah Mahdi ,Dr.Jamal M.Ali* and Dr.Ameer A.Abdul-Rahman

University of Technology Department of chemical Engineering, Iraq

***80002@uotechnology.edu.iq**

Abstract

Hydrodynamic of liquid,-solid fluidized bed was studied experimentally as well as theoretically.The impact of the superficial velocities of liquid, particle diameter ,the liquid holdup, solid holdup, height of bed and minimum velocity fluidizing were investigated. Individual phase holdups were determined through static pressure approach from the information of the pressure drop, the bed height and the properties of the solid and liquid.

Heat transfer coefficients to a liquid-solid fluidized bed in a cylindrical tube have been measured using water as liquid phase and three sizes of spherical glass beads particles as solid phase and electrically heated single vertical tube heated electrically with different heat transfer rate (500, 600 and 700W). The independent variables included heat flux, liquid velocity and particle physical properties. The results showed that the heat transfer coefficient increased with increasing fluidizing water velocity and also with increasing heat flux. It was also determined that the heat transfer coefficient increased with increasing the diameter of solid particles and decreased with increasing the voidage fraction. On the other-hand, the heat transfer coefficient changed with the change of positions of heater distance from the distributer. The experimental results as well as a data bank containing a large number of measured heat transfer coefficients for solid- liquid fluidization over a wide range of operational parameter have been compared with the predictions of most published correlations. The results were agreed in compare with the existing correlations predictions. These correlations were found to be approximate in reproducing the data of present work.

Key word: liquid-solid, fluidized bed, minimum fluidization velocity- solid hold up, heat transfer

AUTHOR INDEX

Ali, Jamal M. **11**
Bradford, Michael **9**
Collins-Chase, Charles **5**
Dicosola, Greg **8**
Dowty, Lauren **5**
Flores, Bernardo **8**
Liu, Shan **6**
Montalbano, Joe **7**
Oh, Hana **1**
Parker, Dan **8**
Schneider, Matt **8**
Sovesky, Robert **4**
Stamato, Howard **10**
Townsend, Paul **5**
Zapadka, Matthew **2**